CORAL reefs are among the most spectacular and diverse natural communities on earth. They are rivalled only by tropical rainforests in the numbers of plants and animals that exist together in a fragile, delicately balanced web of life.

Diving on a coral reef the observer is greeted by a myriad of corals, brightly coloured fish, beautiful shells and a spectacular array of animals and plants of all shapes and colours. The architect of much of this wonder is the tiny **coral polyp** which builds a skeleton of calcium carbonate, and together with innumerable other polyps constructs a reef which provides a place to live for the other organisms in these ecosystems.

Coral reefs in **Western Australia,** although not as extensive as the Great Barrier Reef, have been acclaimed as some of the most beautiful coral formations in the world. In contrast to the Great Barrier Reef, which occurs up to 200 kilometres from the mainland, the relative accessibility of many Western Australian coral reefs allows the wonders of the coral reef to be experienced easily by many people. However, it also makes these reefs more vulnerable to disturbance by human activities.

Left: A great diversity of coral and fish can be seen in the lagoons at Coral Bay, Ningaloo Marine Park.
Photograph: C Bryce.

Left: Spectacular vertical drop-off on the seaward reef slope at Mermaid Reef, Rowley Shoals.
The water surrounding this atoll is about 440 m deep.
Photograph: D Totty.

Above: Many corals can be exposed for several hours at low tides without harm, by secreting mucus to prevent the polyps drying out.
Mermaid Reef, Rowley Shoals.
Photograph: P Baker.

Left: Brightly coloured fish and corals are characteristic of tropical coral reefs. Mermaid Reef. Rowley Shoals.
Photograph: P Baker.

Where Coral Reefs Grow

IN general, coral reefs occur in **tropical seas** and do not exist where ocean temperatures fall below 18° C for extended periods. This may be because most corals cannot reproduce successfully below this temperature, or because juvenile corals are unable to compete successfully for space and light with the abundant seaweeds **(macroalgae)** that thrive in the cooler waters surrounding **temperate** reefs.

In Western Australia extensive coral reefs occur from the northernmost tropical regions in the State (latitude 12° S) to the temperate reefs at the **Houtman Abrolhos Islands** (latitude 28° — 29° S). In most parts of the world, sea temperatures at these high latitudes are too cold for reef growth and the rich coral communities at the Abrolhos Islands probably owe their existence to the **Leeuwin Current.** This warm, tropical current begins to flow southward at about the same time as corals breed on Western Australian reefs, and hence may carry coral larvae south along the coastline during autumn and winter. The possibility that the 'ancestors' of the corals at the Abrolhos Islands were carried south by a current was first suggested by the English naturalist W. Saville-Kent nearly 100 years ago.

Although there are no extensive coral **reefs** south of the Abrolhos Islands there are many coral **species** which grow on limestone reefs. The numerous tropical fish species at Rottnest Island, and the many corals found in the Marmion Marine Park, Geographe Bay and on the south coast near Albany and Esperance, are possibly further evidence of the influence the Leeuwin Current has on the distribution of tropical organisms along the Western Australian coastline.

A coral garden at the Abrolhos Islands. Coral formations in these islands have been described as among the most beautiful in the world.
Photograph: R Lethbridge.

Left: Aerial photograph of Clerke Reef, Rowley Shoals. Because of their isolation these shelf atolls are among the most unspoiled in the world. Photograph: P Baker.

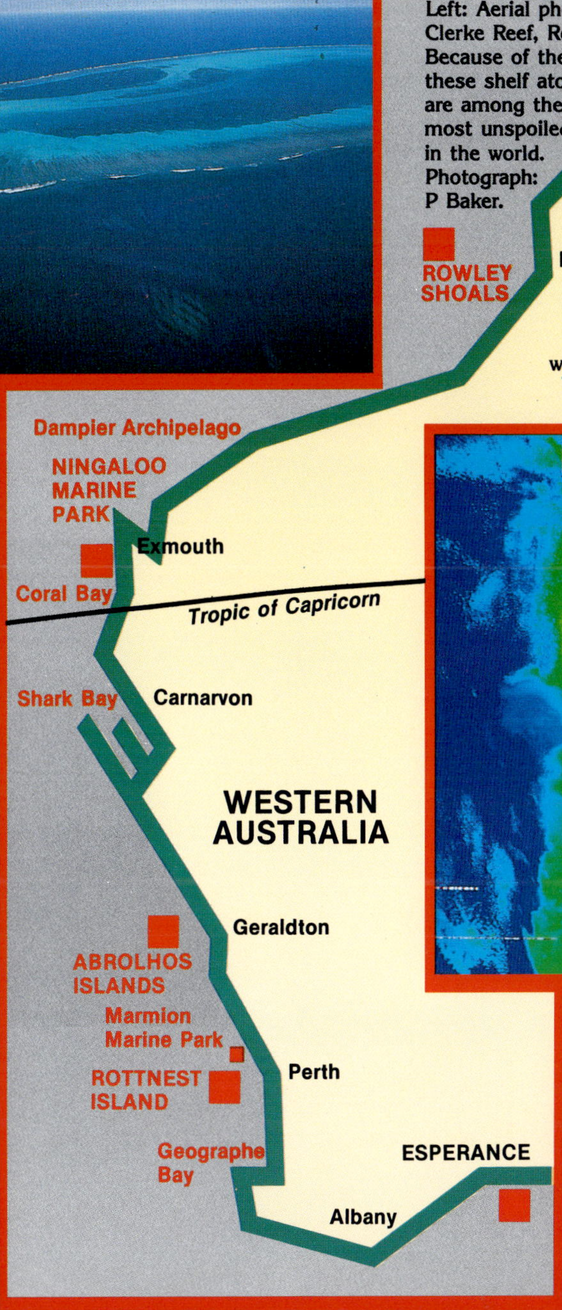

Below: During autumn and winter the Leeuwin Current flows down the coastline of Western Australia and may carry coral larvae to reefs further south. This satellite image shows the flow past North West Cape to the Abrolhos Islands. Warm water is red, cooler water is blue. Photograph courtesy of A Pearce and Curtin University.

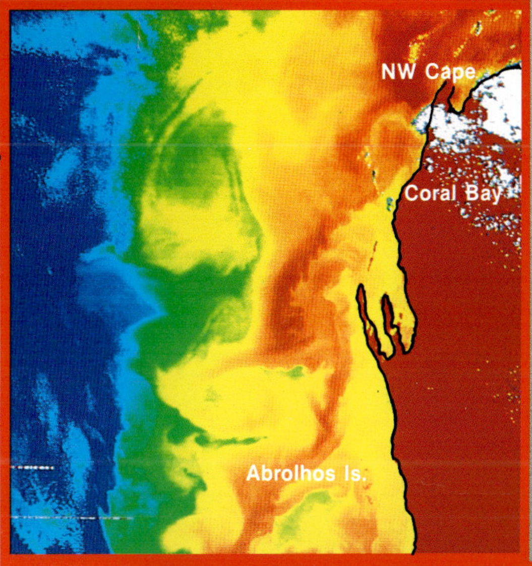

Above left: Ningaloo Reef, in Ningaloo Marine Park, is the largest fringing coral reef in Australia. Its closeness to the mainland is partly due to the generally clear coastal water. Photograph: R May.

Left: The Abrolhos Islands are the southernmost coral reefs in the Indian Ocean. Half Moon Reef and Pelsart Island are shown. Photograph: P Playford.

Left: Many corals occur at Rottnest Island. Pocillopora Reef (seen here), is named after the coral *Pocillopora damicornis.* Photograph: G Saueracker.

Right: The vase coral *Turbinaria* occurs on tropical and temperate reefs, and forms large colonies on the south coast of Western Australia. Esperance. Photograph: G Saueracker.

What Makes a Coral Reef

CORAL reefs are formed by the accumulation and cementation of countless coral skeletons over thousands of years. These massive structures, which can be over one kilometre thick and over a thousand kilometres long, are built by coral polyps which are related to the more familiar sea anemones.

A coral polyp is an animal essentially consisting of a mouth surrounded by numerous tentacles and a simple body cavity. These polyps live in a mutually beneficial **(symbiotic)** relationship with numerous tiny, single-celled plants (algae) called **zooxanthellae.** Corals that contain zooxanthellae are termed **'hermatypic'** or 'reef-building', and the brownish green colour of most corals results from these algae.

Corals obtain most of their food requirements from the zooxanthellae through the process of **photosynthesis.** The balance of their food is provided by a diet of small animals **(zooplankton)** captured by the polyps' tentacles. These are extended at night and contain stinging cells **(nematocysts)** which paralyse the prey.

Staghorn, plate and vase corals at Coral Bay, Ningaloo Marine Park.
Photograph: J Cary.

Mermaid reef at the Rowley Shoals is noted for its steep drop-offs.
Photograph: P Baker.

A tidal passage through Clerke Reef, Rowley Shoals. Currents can exceed 9 knots in these channels. Photograph: P Baker.

Unlike most corals, the polyps of *Goniopora* are expanded during daylight. Photograph: C Simpson.

Diving on a coral reef at Coral Bay, Ningaloo Marine Park. Photograph: Courtesy of the Western Mail.

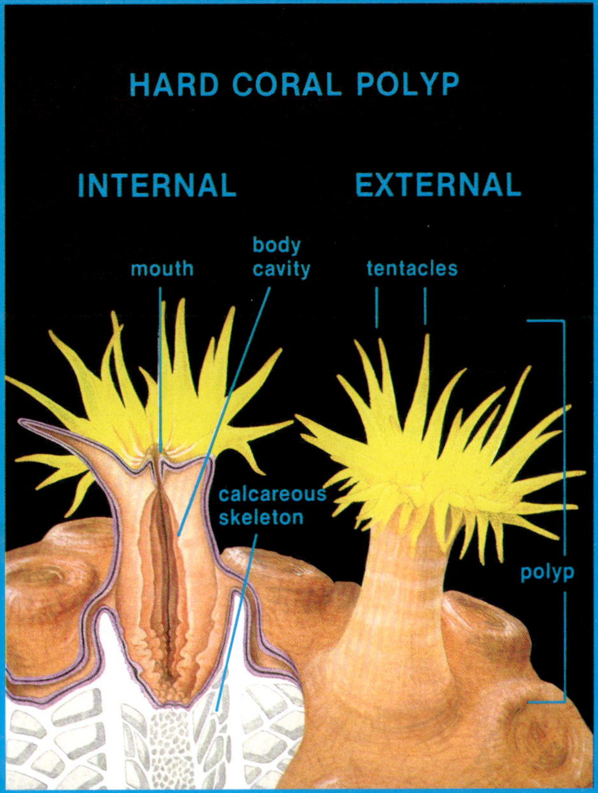

A colony of staghorn coral (*Acropora*), at Ningaloo Reef. Photograph: J Cary.

Growth of Corals

Sea fans or gorgonian corals are not 'reef-building' corals and do not contain zooxanthellae. These animals usually grow at right angles to the current to catch plankton carried past. *Melithaea.* Dampier Archipelago. Photograph: C Simpson.

CORALS grow by budding or cloning new polyps. The many different growth forms found on a reef depend on the species present and on the environmental conditions. For example, fragile **staghorn corals** are found in sheltered lagoons, whereas the low-profile **tabular** forms, that are able to withstand the force of waves, occur on exposed reef fronts. On deeper parts of the reef, sprawling **plate-like** forms maximise capture of the limited available light.

Corals grow at different rates, depending on the species and on the prevailing environmental conditions. Light, seawater temperature and sedimentation are all important influences on coral growth. For example, one species of staghorn coral in the Dampier Archipelago extends its branches by about 140 millimetres a year whereas the same species in the cooler waters of the Abrolhos Islands only grows about 40 millimetres a year. Under ideal conditions many brain corals grow vertically between 10-20 millimetres a year with slower growth in less favourable conditions.

Massive corals (*Porites*) over 4 metres high occur at Ningaloo Reef and the Abrolhos Islands. These colonies may be over 500 years old and therefore could have been alive long before Europeans came to Australia.

A sponge and a brain coral, *Platygyra,* in the Dampier Archipelago. Photograph: C Simpson.

The hydrozoan coral, *Millepora platyphylla*, or 'fire coral', is not a true coral. This species can cause a burning sensation to sensitive skin. Dampier Archipelago.
Photograph: C Simpson.

A starfish clings to a coral at North Bank, off Fremantle.
Photograph: J Cary.

Fast growing staghorn and plate corals (*Acropora*) often dominate shallow reefs leaving little space for other types of coral. Abrolhos Islands.
Photograph: C Simpson.

Soft corals are similar to reef-building corals, apart from the absence of a hard skeleton. Many soft corals contain large quantities of chemicals called terpenes which are either toxic or distasteful to fish. *Sarcophyton.* Rowley Shoals.
Photograph: R Lethbridge.

Above: Nudibranchs are among the most exquisite animals on a coral reef, and some species feed on corals and sponges. *Hexabranchus sanguineus*, Dampier Archipelago. Photograph: C Simpson.

Left: A pincushion urchin (*Asthenosoma varium*) photographed at night at the Abrolhos Islands. The spines can penetrate thick gloves. Photograph: C Simpson.

Two featherstars, or sea lilies (*Comanthus bennetti*), attached to a coral colony. These plankton feeders are related to starfish and sea urchins. Photograph: C Bryce.

Many of the small colourful reef fish, such as the butterfly fish (*Chaetodon*), are obligate coral feeders. If the corals die these fish abandon the reef. Photograph: G Allen.

Pelagic fish such as trevally (*Caranx*) are often found in the deeper water near the reef edge. Photograph: G Allen.

Animals Found on Coral Reefs

Right: The lionfish (*Pterois*) is one of the most unusual fish on a coral reef and is often found in caves or under ledges. Photograph: G Allen.

White tip reef sharks (*Triaenodon obesus*) are found in caves and under ledges of tropical coral reefs. If left alone they are generally placid animals. Photograph: R Beilby.

The clam shown here, *Tridacna maxima*, is common on many Western Australian coral reefs, whereas the giant clam, *Tridacna gigas*, has only been recorded at the Rowley Shoals and at Scott Reef. Photograph: C Simpson.

A rock lobster sheltering in a colony of the coral *Pachyseris*, in 35 m depth at the Abrolhos Islands. The reefs of these islands support an intensive rock lobster fishery. Photograph: C Simpson.

Life Cycle of Corals

MOST corals are **hermaphrodites,** meaning that each single polyp of a colony contains both eggs and sperm.
These hermaphroditic species include most of the staghorn and table corals (Family: Acroporidae) and the brain corals (Family: Faviidae). Other corals are **dioecious,** that is, a colony is either male (polyps contain only sperm) or female (polyps contain only eggs). This group includes the mushroom corals (Family: Fungiidae) and the large coral 'bommies' (Family: Poritidae).

In some corals the eggs are fertilised inside the body cavity, where they develop into **planula larvae** before release.

However, more species release eggs and sperm into the water for fertilisation to take place there. The eggs of many staghorn and table corals develop over about nine months, while the brain corals take only about five months. The sperm is produced over the six weeks prior to spawning. After spawning the eggs are fertilised by the sperm from a different colony of the same species, and gradually change into planula larvae over the next 36 hours. The planulae drift with the currents for another 4-5 days and, if they are still in shallow water, begin to search for places to settle on a reef. If not, they continue to drift and may eventually settle on reefs many hundreds of kilometres away. Very few planulae settle successfully. Those that do, begin to build a skeleton and develop gradually into juvenile corals by budding off new polyps. After several years these corals will be **sexually mature** and able to begin the life cycle over again.

Gonads of the hermaphroditic brain coral *Platygyra sinensis,* showing orange eggs (female) and white testes (male) in each polyp. Photograph: C Simpson.

Eggs and sperm are rolled into a bundle and moved from the body cavity to the polyp mouth in readiness for release.
Abrolhos Islands.
Photograph: C Simpson.

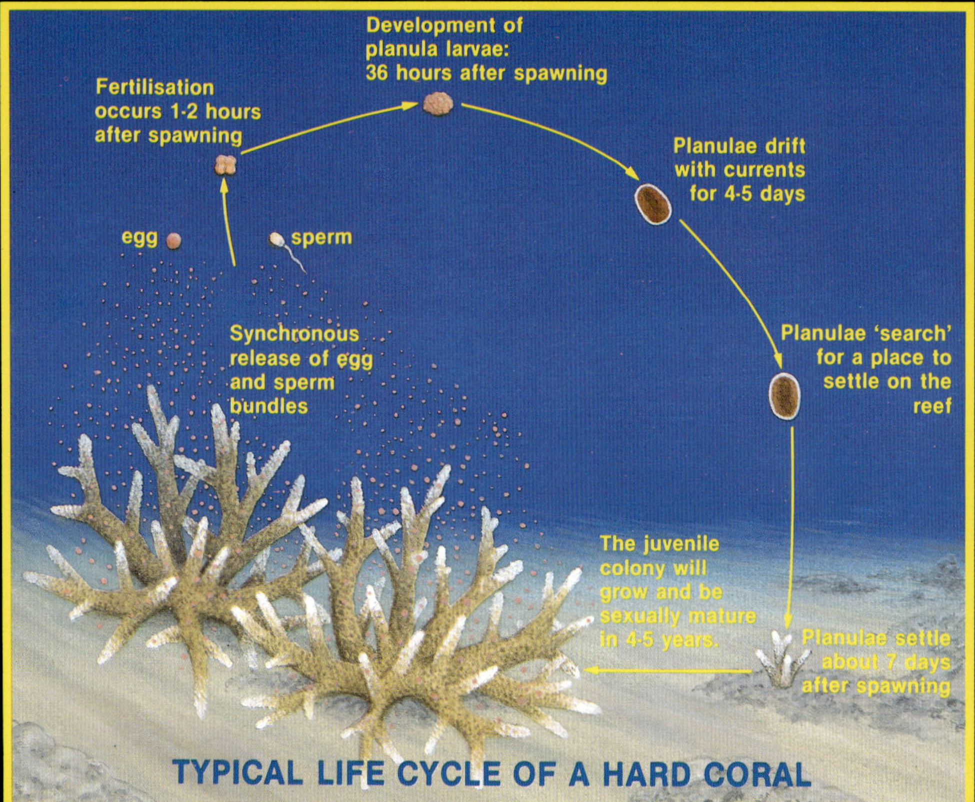

A coral garden at Coral Bay on the Ningaloo Reef showing many juvenile *Acropora* colonies.
Photograph: C Simpson.

Spawning of a mushroom coral (*Fungia fungites*) at the Dampier Archipelago. These corals are solitary (a single polyp) and either male or female. Photograph: W Rowlands.

Mass Spawning of Corals

UNTIL a few years ago most corals were thought to release planulae intermittently throughout the year. Recent studies have shown this to be incorrect and now more coral species are known to spawn at the same time during a brief, predictable period each year. This occurs in spring on the Great Barrier Reef and in autumn on Western Australian coral reefs, and has not yet been recorded elsewhere.

This **mass spawning** of corals is one of the most spectacular events in the yearly calendar of a coral reef. On one or two nights each year, many species of coral suddenly release millions of mainly bright pink egg and sperm bundles which float to the surface of the water. In Western Australia it occurs usually around 8-9 nights after the full moon in March on tropical reefs, and two nights later on the temperate reefs at the Abrolhos Islands. During some years the spawning may be 'split' and occur after consecutive full moons in March and April.

Mass spawning is a mechanism to ensure successful reproduction. By spawning at the same time, colonies of one species will maximise fertilisation while the simultaneous spawning of many different species will ensure enough planulae survive by swamping predators with an excess of food over a short period. Similarly, spawning at night rather than during the day may be a clever way to minimise predation by animals which need light to see their prey.

The timing of coral spawning in Western Australia coincides approximately with the beginning of the flow of the Leeuwin Current, and planulae may be swept southward to supply reefs further south with new corals. Thus, many coral reefs along the Western Australian coastline may be biologically connected and the long term survival of southern coral reefs, such as at the Abrolhos Islands, may be partially dependent on the health of reefs further north; for example, the Ningaloo Reef.

As the sun sets at Ningaloo Reef, divers prepare to go below and watch the corals spawning.
Photograph: Courtesy of the Western Mail.

Diver watches a coral colony release millions of egg and sperm bundles into the water.
Photograph: Courtesy of the Western Mail.

Spawning of a *Galaxea* colony at the Dampier Archipelago.
Photograph: C Simpson.

If the weather is calm after spawning, coral eggs and larvae can form pink slicks several hundreds of metres long. Coral Bay, Ningaloo Reef. These slicks can be confused with red tide blooms, which are planktonic algae. Photograph: K McAlpine.

The common table coral, *Acropora spicifera*, at the Abrolhos Islands, showing the large quantities of egg and sperm bundles being released during spawning. Photograph: C Simpson.

Coral Predators

LIKE most animals, corals have natural enemies. By far the best known predator of corals is the **crown-of-thorns starfish** (*Acanthaster planci*) which, over the past 30 years has damaged extensive areas of living coral on the Great Barrier Reef.
This starfish is relatively common in the Dampier Archipelago, but is only rarely sighted at other coral reefs in Western Australia, and has never been recorded at the Abrolhos Islands. The reason for the apparent low densities of this starfish on Western Australian coral reefs, other than in the Dampier Archipelago, is not known. A possible explanation is that water circulation patterns at the time the starfish spawn (possibly in mid-summer) prevent the widespread dispersal of larvae of this coral predator.

Another common predator of corals is a **marine snail** *(Drupella cornus)* which has caused considerable damage to coral reefs in the Pacific and to reefs off Japan. *Drupella* is common on Ningaloo Reef and appears to be causing extensive damage to live corals. The effects of this predation on the ecology of Ningaloo Reef are unknown, but the snails have reduced formerly attractive areas of live coral, with numerous colourful fish, to coral rubble encrusted by algae with few fish.

Fish also prey intensively on some species of live coral; **parrot fish** for example, have teeth adapted for biting off pieces of coral or for scraping off polyps.

Parrot fish (*Scarus*) bites on *Porites* 'bommies' in the Dampier Archipelago.
Photograph: C Simpson.

Crown-of-thorns starfish feed preferentially on *Acropora*. Dampier Archipelago.
Photograph: C Simpson.

The crown-of-thorns starfish, *Acanthaster planci*, feeds extensively on live corals by everting its stomach and digesting the polyps, leaving the skeleton intact.

The marine snail, *Drupella cornus* (see arrows), appears to feed preferentially on the soft tips of *Acropora* branches, leaving the corals with a characteristic 'chiselled off' appearance that soon becomes encrusted with algae. Coral Bay, Ningaloo Reef.
Photograph: M Forde.

Feeding scars (white area) on an *Acropora hyacinthus* colony caused by the crown-of-thorns starfish. The dead skeleton is rapidly covered by algal growth. Dampier Archipelago.
Photograph: C Simpson.

Why Manage Coral Reefs

Potato cod (*Epinephelus tukula*) are extremely inquisitive and readily approach divers. Because they succumb easily to fishing pressure, they are usually only found around reefs rarely visited by people. Photograph: D Totty.

Far right: Diver tagging staghorn coral for reproductive study. Scientific research is necessary for effective management. Coral Bay, Ningaloo Reef. Photograph: M Forde.

Right: Green turtles (*Chelonia mydas*) are commonly seen around coral reefs and come ashore to nest on beaches. Constant disturbance by people causes turtles to abandon traditional nesting grounds for more secluded sites. Dampier Archipelago. Photograph: K Morris.

CORAL reefs are extremely complex and delicately balanced communities. Natural disturbances such as cyclones can devastate extensive areas of reef which take decades to recover fully. These events, however, are some of the natural processes shaping a reef which evolves over many thousands of years. A living coral reef is a three dimensional structure providing living space, food and shelter within the reef framework for many different organisms. These organisms attract other animals which feed on them, and are part of a long and complex food chain that starts with the reef itself. When corals die they are encrusted with plants **(algae)** and attacked by boring organisms that gradually weaken their skeletons. Eventually the honeycomb nature of the reef can be lost if the reef framework collapses. Many small animals and fish which lived within this framework, and their predators, would then depart leaving a barren 'desert' of coral rubble.

Overfishing of **territorial reef fish,** such as coral trout, and intensive collecting of live shells can deplete these resources so rapidly that a major attraction may be lost in a few years. **Anchor damage** or **reef trampling** can reduce formerly attractive reefs to unsightly areas devoid of their original beauty. Activities associated with industry (oil spills and dredging) or tourist developments (sewage discharge) are also potentially very harmful to coral reefs.

Very little is known about the relationships between different components of a coral reef and the wider implications of a particular impact may not be realised until it is too late. The depletion of a particular fish species for example, may cause a population explosion in its prey which may then, in turn, become a problem. Similarly, chronic low levels of a pollutant may not kill organisms but may seriously affect their reproductive cycles. Thus, coral reef managers have to be extremely cautious when considering potential impacts on a coral reef.

The management of coral reefs in Western Australia, and management of our environment in general, depends largely on the goodwill and common sense of the public. Most activities in isolation have only a minor impact on the environment. However, many activities are widespread and must be viewed from this perspective if valuable natural resources such as the coral reefs of Western Australia are to be conserved, and hence kept available for the appreciation and use of future generations.